EARTH IS A SEEDBED

EARTH IS A SEEDBED

poems
by
LOIS O'CONNOR

REVISED & EXPANDED SECOND EDITION

THE SCRIPTORIUM PRESS
Alfred, New York
1986

EARTH IS A SEEDBED
Poems by Lois O'Connor

Revised and Expanded Second Edition

Edited and designed by Diane Martin

Published by: THE SCRIPTORIUM PRESS
71 South Main Street
Alfred, New York 14802

First Edition: Tauvirg Publications
Freeville, New York
1974

Printed and bound by The Sun Publishing Company, Alfred, New York.

ISBN 0-931485-26-6
Library of Congress Catalog Card Number 86-063215

TABLE OF CONTENTS

EARTH
IS A
SEEDBED

SEEDBED

Earth is a seedbed
nurturing root and mineral,
the feathered and the furred,
the human soul.

Goat and goldfinch,
crystal, clover,
mollusk or man -
whatever kingdom -
each must grow
to a fulfillment ...
and a sacrifice.

WIND OF CHANGE

A wind of change
disturbs the Cosmos,
lashing the Ocean of Space
moon-raised to rip tide.
Floodgates of the Piscean Age
give, as torrents
of Aquarius spin
the tidewheel of the Universe.

COAST-BRED

The soughing of the autumn wind
through ash and oak (an inland surf-sound)
laps at coast-tuned ears.
Less strident than the cry of ocean gull
rises the scream of jays
in gossip with their migratory kin.
Stout hemlocks bound the inland distances,
but, in the marrow of emotion,
never substitute for ribs
of coastal rock.

DRIFTWOOD

Fragment of driftwood
salt-silvered by the wash of tide,
shaped by primordial power,
within the simple contour of your side
has moved the force
that raised an Everest
and crashed the walls of Jericho.

THE PEBBLE GATHERERS

From rise of morning sun
to evening afterglow
the pebble gatherers
search the sandy slopes.
Shell not fossil catch
their beach-bound eyes.
Pebbles only pebbles
are the treasure
carried off for winter solace.
They stoop again
and yet again to gather up
the bits of colored stone
that spell romance:
particles that punctuate
their days.

WIND-SUNG

Wind-sung
the meadow sleeps:
No leaf, no petal stirring;
hushed the cricket serenade.

Wind-sung
the stars awake:
Spangle a black-breasted sky,
dance to flute and harp of wind.

Moon-sung
the Cosmos stirs:
Binds with all-pervading light
silver star and clover.

PERSPECTIVE

The seasons in a circle -
spring daffodil to saffron autumn crocus
the paintings hung -
an artist's garden on a gallery wall.

Jacinth to tangerine, tiger lillies
dominate one sun-struck study;
cathedral-window colors glow in iris
cross-hatched by shafts of indigo.

In autumn tapestry
damascened with jasper and with jade
brown pods of milkweed swing
beside a chalky skeleton of Queen Anne's lace.

Fashioned with ink, casein and artist's brush
the paintings offer me a season's circle
suited to the span of my perspective.

Ringed by no time-space moat
The Celestial Eye
girdles the living circle
in omniscient glance.

OMENS OF SPRING

I

Limned on the backdrop
of a moon-blanched
February sky,
the crow,
a solitary smudge.

II

Starlings,
wind-roughed feathers
puffed against the cold,
huddle deep among
snow-bent hemlock boughs.
From root to frozen branch
does some warm sentiency
of spring vibrate comfort
to an icy claw?

MONTH OF THE CROW

Some sentiency of spring
animates the torpid crow
humped against
a February dawn.
Does foreknowledge ride the wind
ruffling his dusky feathers
or is he roused
by vibrant branch
of sap-invigorated maple
to sound his raucus joy?

MICROCOSM

Circumference of a spring day
closes around a lonely stretch of land.
Unharried deer and rabbit,
bluebird and nesting robin
move unafraid in morning peace,
where footfall seldom raises
dust or human sound
along the abandoned wagon road.
Mid-day sun lances deep through
tangled wands of flowering almond
to a blackened hearthstone,
raising ghosts of embered warmth
in the crumbling chimney's
light-flicked maw.
Moonlight slips from bough
to gnarled bough of unpruned apple tree,
seeps into crannies of a stone wall
bereft of home-bearing.

Lilac fragrance,
a breath of continuity,
laces the isolation.

NOW IN APRIL...

In November
the old dog turned
to face his winter.
The snows
that gently quilted
his earth-sheeted bed
were no whiter
than his silvered muzzle.

Now in April
daffodils
no longer bend
to spaniel paw.

PETAL-PINK ILLUSION

One petal-falling afternoon
a path among the orchard lanes
led to enchantment of a half-moon night.

Spring's promise fruited on the bough.
Love's image laid upon the heart
was born of petal-pink illusion.

PINK IS JUNE

Pink is the color of June:
Blush of dawn's rose-gathering clouds
emerging from night-sky cocoon,

Rose of mallow tufts the hedge,
ribbons through crimsoning berries
and Queen Anne's lace at meadow's edge.

Clove pink spices sapphire air.
Pink of Damask, China, Moss -
June-fragrant roses everywhere!

Pink is the color of June
affirmed in sweeps of coral sky,
symbol in summer's ancient rune.

SUMMER DAY

Sunup
to moonrise, all
a daisy-eyed summer day,
the meadow sang with bees searching
honey.

REFRAIN

Muffled
against the summer-scented night,
note of a cowbell,
an old song from the meadow.
Contentment
grows with the dusk.

DAY'S JOURNEY

The journey began quietly
along a road defined
by mallow pink and chicory blue;
the scent of milkweed
weighting morning freshness
toward a mid-day lethargy.
From this initial arc
what premonition
could instantly round out
a summer's day circumference
with foresight of the journey's end
along a roadway strewn
with swaths of orange lillies
prostrate by a mower's whim.

ADMONITION

Gather
dill for witchcraft,
hyssop and mandrake root;
for the daft, a seed of peony
to brew.

Bundle
sprig, root and twig -
milkweed for warts, boneset -
best leave unplucked the bitter leaf
of rue.

VARIATIONS ON A THEME

Sun draws radiance
from white-petaled peonies
splashed with morning dew.

A gift of summer
peonies flaunt their beauty
by the garden wall.

Peony petals
delight the eye with color
touched by summer light.

When summer days end
dusk enfolds the peonies.
Fireflies and silence

NURTURING

Her plants
are tucked about
in unlikely corners,
with care, coaxed from barrenness to
beauty.
Her love
sheds comfort to dark crannies of
human needs; as with plants,
nurtures lives
to fragrance.

GARDEN MUSIC

A song
flows through my garden -
cadence in color -
the refrain is cosmos.

LEPRECHAUN

A leprechaun lives
In my garden wall
In a crooked cranny
Bigger than small.

He's a mischievous creature
Half-high tall.
Sometimes I doubt
He's there at all.

PIONEER MOTHER

Tell the hours bindweed to sow thistle
in nightmare fear.
Pray.
Steep bloodroot for choking cough;
dill to sooth the fevered child.
Tell the hours primrose to marigold.
Spell fatigue with boneset, ease the heart
with faith's hyssop.

JOE PYEWEED

Joe Pye
if you could see
the purple-blooming weed
that bears your name,
the weed,
from which, with Indian wisdom,
you extracted simples
to ease your neighbors' ills:
If you could know the joy
its purple, roadside splendor gives,
Joe Pye,
it would bring
a gleam of pleasure
to your eye.

SUMMER INTERLUDE

Cat, Snake
Wren and I,
warmed by the sun -
no more than a teacup full -
held conversation
by the garden wall.

Comfort-seeking Cat
curled, straightway,
on the nearby chaise
from there to raise
his questions.

Snake,
on sun-warmed stone,
uncoiled, but ever wary,
made wise comment
with forked tongue.

From vantage
in a yearling sumac
Wren showered opinions
for all to heed.

Cat, Snake, Wren -
what human words
could match
their disparate commentary,
essence from some Ancient Wisdom?

No words!
Only love,
transcending speech,
flowed through sapphire filaments
of summer light.

CHOICE

The rover in me longs to go
in carefree, featherweight delight
with every dandelion blow.

It is the heart seeks to transmute
the earth-warm love and sun-bred strength
that gives the dandelion root.

AUGUST
(For MEW)

August,
Virgo-nurtured,
coaxes and cudgels
toward perfection
the ripening harvest
augured by full flowered spring.

Sinew of intuition
binds the August days
to burnish copper-lustre grain,
enrich the golden peach,
thicken brown cattails,
plump the root of ginseng.

Each hour,
Virgo-disciplined,
is bent to consumation
of the springtide promise
before the muted sigh of Libra
tips the cycle's balance.

PREMONITION

I shall not live past time
when seeds grow heavy in the headed grain,
when kernels swell the corn husks green
and crimson steals along the apple's side.
I shall not live
for promise of fruition
stirs not in the body's husk.
Spring left no seed to swell the spirit
as it did to bring the apple into bloom,
the clover and all fruits of the earth.
The core of self requires
a seedtime and a harvest for its growth,
else are the sun, the wind, the rain of being
wasted on infertile ground.
I shall not live past time
when seeds turn jet within the melon's heart,
when milkweed floats in burnished air
and harvest's magic rings the earth.

AUTUMN PRELUDE

Fall came early to the swamp.
No Indian Summer softness
bridged first chill frost
to winter-honed days.
No remnant summer sun
brought lingering warmth
to small turtles huddled
on moss-grown logs sunk
half-way in the muck.

Fringes of frosted bracken
now bind a rim of bronze
at water's edge.
Raw wind riffles tussocks of rush
that shelter muskrats.

Wind of premonition
invades the brittle bone;
Autumn binds the days.

SEPTEMBER SUN

September's mellowed sun
transmutes from solar strength
a tenuous compassion to bank
the fire of summer
against a winter's need
for embered warmth.

SEPTEMBER

September's speech
is simple and direct,
phrased in singing arc
of goldfinch flight
rhymed with flicker-knock
on hemlock tree.

September's voice
is harvest-fragrant wind
filtering the woods
to set leaf-palms
of poplar clapping
for an antic squirrel.

OCTOBER

October brims the cask of winter
with a heady brew -
compound of aster,
marigold
and meadow rue,
distilled in juice of apple
quince,
wild plum and pear;
incarnadined with grape;
filtered through nectared air.

Each flagon drawn in snow-chilled hours
holds a memory-draught in store -
honey-sweet and bitter-sweet -
October's evanescent core.

SNOW FALLS LIGHTLY

Snow falls lightly on the stark earth,
comforting the barren soil
with tender touch.
Heart wounds are soonest healed
by love that comes
silently as snow.

MUSKRAT WINTER

Muskrat build snugly
under alder and willow.
House grounded neatly
on tussock of rushes,
icebound, will protect you
from fox, hawk and weasel.

Spread a leaf carpet,
Fresh moss for a pillow.
Choose a bog cupboard
with choicest of provender.
Supper on sweet flag,
clam, yellow lily.

While winter bustles
through seer cattail and willow,
Muskrat, sleep soundly.
Wait content for sun's warmth
to bring life to marsh marigold,
clover-green to the meadow.

THE PREDATOR

Love takes its meat,
sucks the marrow from the bones
each one cleaned white and bare.
Fed to satiety, the predator discards
the stripped and desiccating frame
expecting it to rise again and dance;
nor cares that it will always move
with clatter of a skeletal marionette.

VALLEY OF WEEPING

"The bitter waters of sorrow"
surge
through caverns of the heart.
Through Baca, Valley of Weeping,
they race
a riverbed of merciless anguish.

Grief, at crest,
may tap
a watercourse of wisdom
whence a sweet stream
of immaculate compassion
flows
to shrive the bitter flood.

RACHMANINOFF RHAPSODY

Filaments of melody
shaped to an enchanted strand:
Fragments of crushed stars
tossed from an angel's careless hand.

WHAT HALF-CHORD ...

What half-chord
taunts the heart;
eludes the net
of memory?

Perception
seeks the integer
in arcane notes
that will release the primal chord
in full-toned vibrance
on the inner ear.

VIGNETTE IN SPAIN

Three white windmills
on a hill
above the silver green
of ancient olive trees -
windmills tilting
at uncertain wind
as Don Quiixote tilted
with a dream.

PROTEST

There is no wholeness to a day.
A spate of duties unessential
strips each firm core
to shreds inconsequential.

THE PRIZE

It was ill-omened,
that morning Paris, boldly
claiming Aphrodite's prize,
seized the radiant Helen.

Predatory, triumphant,
Paris set the sails for Troy:
Exhilaration shook the air.

In somber rhythm with the sea
Helen's poignant thredony
rose and fell
until a spark from Paris' mood
burned a flush on her pale face
and alien rapture laced her grief.

A shudder swept her every bone and sinew,
portent of the shattering battlements of Troy.

SPARTA REMEMBERED

It was a golden world
that morning Helen dallied
in a Spartan meadow
gathering elecampane.

Apollo's blazing chariot
drew a golden arc across the sky,
the shining splendor
repeating in the golden ribbon
binding Helen's hair;
reflecting in the yellow-petaled
elecampane.

It was a star-crossed world
that morning, Paris boldly
claiming Aphrodite's prize,
seized the radiant Helen.

Blood red the world
when Helen, shivering on the battlements,
saw carnage on the Trojan plain,
watched her husband, Menelaus,
exact the final dreg of vengeance.
Did her heart quicken, her beauty kindle,
to half-remembered lineaments of home
and scattered stalks of elecampane.

SUPPLICATION

Gethsemane
is a heritage to every soul.
Shape us to sustain the grief:
Will chastened but unbroken by the pain;
Courage, a vigil flame
to warm us numb in agony;
Faith, that from the fire of suffering
the spirit, crucified, will rise.

QUESTION

What ancient lineage
lingers in the hawk
poised erect
atop the weathered oak?
His hooded eyes hold
memories, still, of Pharaoh's land
his beak, like Horus,
in ageless silhouette.

HANDS ARE:

Cruelty runs deep.
For murder, child abuse, rape
hands are instruments.

Formed by heritage,
shaped by circumstance
hands are revealers.

The flesh, heart, spirit
are renewed by human touch.
Hands are the healers.

From the heart's longing
love consumates desire.
Hands are translators.

VICIOUS CIRCLE

Stiletto sharp
the needle point thrust
to the heart of a French knot -
a vicious stab!
Sewing circle, circle, circle,
Elocution
Execution!

Linen, hoop-noosed,
taut and stretched for stitches
delicate with daisies -
a pretty sight!
Sewing circle, circle, circle,
Elocution
Execution!

Featherstitch imagination,
needle-tongued assassination.

WHERE TIME SINGS

Sunup to moonset
where love is
Time sings.
Notes of wind and water,
woods, wetlands,
wings and wildflowers
coalesce to chords of joy.

DAY OF DELIGHT

Today,
a round, fresh fruit,
was tossed
for me to catch.
At cycle's end
it had become
a pomander, firm and spiced,
hung in the heart
to scent less fragrant days.

ACKNOWLEDGMENTS

Appreciative acknowledgment is made to the following publications in which some of the poems in this collection first appeared: The American Agriculturist, The American Bard, American Weave, Cornell Plantations Quarterly, The Country Poet, The Rochester Democrat & Chronicle, The Hartford Times, The Herald Tribune, Herb Grower Magazine, Highlights for Children, Imprints Quarterly, Ithaca Grapevine, Ithaca Times, Lyric Leaves, Maine Life, New Athenaeum, Oak Leaves, The Stone #1 & #2, and three anthologies, "Torn Out of Time," "Turn the Page" and "In Spite of the Dragon."

OTHER PUBLICATIONS BY LOIS O'CONNOR

Crossroads Comment, Volume I
The Ithaca Journal
Ithaca, New York
1951

Crossroads Comment, Volume II
The Ithaca Journal
Ithaca, New York
1953

A Finger Lakes Odyssey
North Country Books
Lakemont, New York
1977

...Of Tarragon, Thyme & Tauvirg, Delectable Cooking with Herbs & Spices
Revised and Expanded Second Edition
The Scriptorium Press
Alfred, New York
1987

COLOPHON

This book was printed in Iris PS typeface on a Canon VP 3000 Word Processor. The cover illustration is from a title page designed by William Morris, Kelmscott Press, 1892; the illustrations are from an ornamental title page from a Psalterium, Cologne, 1518. It is printed on acid-free Hammermill paper.